The 60-Minute Quantum Physics Book

Science Made Easy For Beginners Without Math and in Plain Simple English

Donald B. Grey

The 60-Minute Quantum Physics Book

© Copyright 2020 - All rights reserved.

The content contained within this book may not be reproduced, duplicated nor transmitted without direct written permission from the author or the publisher.

Under no circumstances will any blame or legal responsibility be held against the publisher, or author, for any damages, reparation, or monetary loss due to the information contained within this book, either directly or indirectly.

Legal Notice:

This book is copyright protected. It is only for personal use. You cannot amend, distribute, sell, use, quote or paraphrase any part, or the content within this book, without the consent of the author or publisher.

Disclaimer Notice:

Please note the information contained within this document is for educational and entertainment

The 60-Minute Quantum Physics Book

purposes only. All effort has been executed to present accurate, up-to-date, reliable, complete information. No warranties of any kind are declared or implied. Readers acknowledge that the author is not engaged in the rendering of legal, financial, medical or professional advice. The content within this book has been derived from various sources. Please consult a licensed professional before attempting any techniques outlined in this book.

By reading this document, the reader agrees that under no circumstances is the author responsible for any losses, direct or indirect, that are incurred as a result of the use of the information contained within this document, including, but not limited to, errors, omissions, or inaccuracies.

The 60-Minute Quantum Physics Book

Bluesource And Friends

This book is brought to you by Bluesource And Friends, a happy book publishing company.

Our motto is **"Happiness Within Pages"**

We promise to deliver amazing value to readers with our books.

We also appreciate honest book reviews from our readers.

Connect with us on our Facebook page www.facebook.com/bluesourceandfriends and stay tuned to our latest book promotions and free giveaways.

The 60-Minute Quantum Physics Book

Table of Contents

Introduction

Chapter 1: Quantum Physics - An Overview

    Quantum Mechanics

    Principles of Quantum Mechanics

    Developing Quantum Theory

    Implications of Quantum Theory

Chapter 2: Why Should You Learn Quantum Physics?

    Quantum Computing

    Cryptography

Chapter 3: String Theory

    Strings and Membranes

    Quantum Gravity

    Unification of Forces

    Supersymmetry

    Extra Dimensions

Chapter 4: Bohr-Einstein Debates

    Quantum Mechanics Conference

    Einstein's Criticisms

The 60-Minute Quantum Physics Book

Chapter 5: Quantum Entanglement

    Quantum Complementarity

Chapter 6: Schrödinger's Cat

    Superposition

    The "Experiment"

    The Many-Worlds Interpretation

Chapter 7: Double-Slit Experiment

    The Value of Uncertainty

Chapter 8: Teleportation

    Is Teleportation Practical?

Chapter 9: Interesting Quantum Physics Facts

    The Entanglement Concept

    Virtual Particles

    The Weight-Speed Argument

    Sounds Like Magic

    The Thrills of Quantum Theory

Conclusion

References

The 60-Minute Quantum Physics Book

Introduction

Quantum physics is an interesting topic. Everything on this Earth exists in some quantum principle. In the simplest terms, quantum physics is the branch of physics that attempts to explain how and why things behave the way they do. It is the branch of science that comes closest to giving us an in-depth understanding of matter by studying the particles that make up matter and the forces that act on the particles to behave the way they do. After all, matter is anything that occupies space and has mass. Matter is all around us.

Like many people, the very mention of "quantum physics" can send chills down your spine. Thinking of all the math involved might have you running for the hills, yet that shouldn't be the case. If anything, a lot of people have learned so much about quantum physics in a fun way - watching the Big Bang Theory! Dr. Sheldon Cooper and his friends did an excellent job at making quantum physics relatable, and in this book, we will do just that - introduce you to quantum physics without all the Einstein-level math.

Quantum physics is an in-depth study of atoms - the

The 60-Minute Quantum Physics Book

smallest unit of matter. Atoms make up everything on Earth. Liquids, solids, gas, atoms make up everything. Quantum physics digs into the workings of atoms, and from this knowledge, we can also understand why biology and chemistry work the way they do. From this explanation, we can tell that, at some level, quantum physics is always at play on everything on Earth.

Perhaps you are trying to explain how solar energy turns into electricity, how computer chips work, what artificial intelligence entails, and so on; quantum physics is always involved, directly or otherwise.

Quantum physics is, perhaps, one of the fields of science that is shrouded in mystery, and might be a paradox in itself. Why do we say this? Well, for starters, there is not a single quantum theory that is acceptable across the board. This means that there is so much to learn and discover in the field of quantum physics. Quantum mechanics, developed by Erwin Schrödinger, Niels Bohr, and Werner Heisenberg, is a framework acceptable to all quantum physicists and scientists.

Discussing quantum mechanics, or Einstein's theory of relativity, among other theories, might be overwhelming for most people. Instead of going into the gritty details, this book gives you a more relaxed

The 60-Minute Quantum Physics Book

approach, such that you can learn quantum physics by understanding some of the key concepts, and then from there, go on to build on complex theories and subjects at your pace.

The 60-Minute Quantum Physics Book

Chapter 1: Quantum Physics - An Overview

Physicists often use quantum physics, quantum theory, and quantum mechanics interchangeably. Ideally, quantum mechanics encompasses all discussions in the quantum realm of physics. Quantum mechanics is a science that studies the behavior of matter and light at atomic and subatomic levels (Squires, 2018). In this study, we try to understand the properties of atoms and molecules, especially how they interact with electromagnetic radiation and other particles in the universe. In physics, mechanics studies movement of anything from microscopic particles to planets.

Quantum Mechanics

Quantum mechanics has proven successful over the years, albeit in theory. Learning about quantum mechanics awakens you to some strange conclusions about the reality of the physical world we live in. For example, taking the fundamental approach to quantum physics, we focus on subatomic particles. Drawing conclusions based on studies of subatomic particles, you will realize that many equations used in classical mechanics might not be useful (Wilce, 2019). These are equations that describe motion at speeds and magnitudes we can see and interpret with the naked eye.

In our daily lives, what we can visualize and conceptualize is that an object can only exist in one place at a specific time - that moment when you interact with it. In quantum mechanics, however, the existence of objects is a matter of probability. Quantum theories suggest that, at any given point in time, there is always a probability that the object might exist at different points. Sounds crazy, right?

The 60-Minute Quantum Physics Book

Matter is made up of atoms - at least we can all agree on that. Atoms are composed of electrons, continually orbiting a nucleus of neutrons and protons. The nature of atoms is that they are discrete, meaning that they can either be in one place or the other. The study of quantum mechanics has evolved over the years, initially starting as controversial experiments to explain phenomena that classical mathematicians or mechanics were unable to.

Principles of Quantum Mechanics

Many scientists contributed to the development of quantum mechanics towards the end of the 19th Century. Early in the 20th Century, Einstein published the *Theory of Relativity*, a mechanical study of motion at high speed (Hiton, 2017). Their work, together with Einstein's study, informed the three principles that have guided quantum mechanics below:

- **Quantized properties**

In quantum mechanics, you learn that some properties of matter can only exist in a specific amount. Examples include: Color, speed, and

The 60-Minute Quantum Physics Book

position. This contravenes classic assumptions in mechanics which suggest that such properties can only exist in a continuous spectrum. In physics, "quantized" means setting a limit for possible values such that the variables in question can only exist within discrete magnitudes.

A quantized property, therefore, means that you can only prescribe discrete values to the properties of matter like position, speed, and color. Using the example of an electric charge, you can only assign integer values to identify electrons. When referring to electrons, you can only use whole number integers, like 3, 6, and 8. You cannot have an eight of an electron, or a half an electron. Other properties of matter that can be quantized include momentum, energy, and mass.

- **Particles of light**

For centuries, scientists have often held that light travels in waves, like the ripples on the surface of a calm water body. However, quantum physicists challenged this thought, suggesting that light behaves like particles.

The 60-Minute Quantum Physics Book

Further to this, light can bend around corners, and bounce off walls. The troughs and crests that make up light waves generally cancel out when they add up. From this explanation, adding more crests to light waves increases brightness, while canceling the crests produces darkness.

We already mentioned that matter is made up of tiny particles (electrons, neutrons, and photos). Similarly, quantum theory supposes that light is also made up of tiny particles (photons). Wave-like properties of light can only be experienced with microscopic particle mass.

- **Matter waves**

For years, scientists proved and held that matter can only exist as particles. However, quantum physicists suggested that matter can also have wave-like properties, becoming a key argument in quantum mechanics. All forms of matter exhibit wave-like tendencies. For example, you can diffract a beam of neutrons just as you would a wave of water or photons. This theory, however, might be impractical because in most cases, the particle wavelength is too small to register in daily activities, and for this reason, matter waves might be irrelevant.

Developing Quantum Theory

Using quantum theory, you can explain the behavior and nature of energy and matter at the subatomic level. This is what we commonly refer to as "quantum mechanics" or "quantum physics". The first assumption of quantum theory was presented by Max Planck in 1900. He observed that, with increasing temperature, the color of radiation from a source of light transforms from red through orange to blue.

From this observation, he supposed that, like matter, energy exists in individual units. Scientists had previously held that energy can only exist as an electromagnetic wave. Planck's assumption, therefore, meant that as we can quantify matter, we can equally quantify energy. He later referred to the individual quantifiable units of energy as "quanta" (discussed under "quantized properties" above).

Planck's work was soon followed by Einstein, who theorized that, apart from the energy, we could also quantize the radiation in a similar manner. Later on, Louis de Broglie suggested that at the atomic and subatomic levels, energy and matter exhibit the same behavior. They can either behave like waves or particles. His observations gave birth to the famous wave-particle duality theory which, in simple terms, states that particles of matter and energy assume wave-like or particle properties under different conditions.

Implications of Quantum Theory

As much as quantum physics poses several challenges in relation to the physical world, we cannot turn a blind eye to the implications therein. Two interpretations have ushered in significant growth in the field over the years:

- **Copenhagen Interpretation**

The 60-Minute Quantum Physics Book

This theory was proposed by Niels Bohr, an assertion that particles can only be identified as what we measure them to be. We cannot, therefore, assume that particles have unique properties, neither can we assume that they exist before we measure them (Forrester, 2018). From this understanding, Bohr's suggestion was that objectivity is non-existent.

It is from this theory that the principle of superposition was formed. Superposition suggests that while we might not know the immediate state of an object, it exists simultaneously in all the states imaginable, as long as we do not check to confirm its existence. We will discuss this concept later in the book.

- **Multiverse Theory**

The 60-Minute Quantum Physics Book

The theory of the multiverse commonly referred to as the "many worlds theory" is another vital interpretation of quantum theory. This theory suggests that immediately an opportunity arises for an object to exist in any state, the universe in which the object exists transforms into several parallel universes equal to the number of states where the object can possibly exist. Each of the transmuted universes holds one possible state of the object under observation.

Apart from that, the multiverse theory also suggests the presence of a mode of interaction across the parallel universes which allows each of the states in all the universes to be accessible, and in this case, any action on the state in the present universe affects the possible states in all the other universes in some way.

For years, scientists have struggled to come to a consensus about different quantum concepts. Popular scientists like Einstein, Planck, and Bohr advanced several proposals about their knowledge of quantum mechanics. Their theories have since been used in further experiments over the years by scientists from different divides - those trying to further their concepts of quantum physics, and those who were out to disprove their theories.

Modern physics as we know it has its foundation in Einstein's theory of relativity. Together with quantum

The 60-Minute Quantum Physics Book

theory, these two concepts have shaped the future of modern physics, and from there, we can look forward to an exciting future in terms of developments in the quantum field, particularly quantum cryptography, computing, chemistry, and quantum optics.

Chapter 2: Why Should You Learn Quantum Physics?

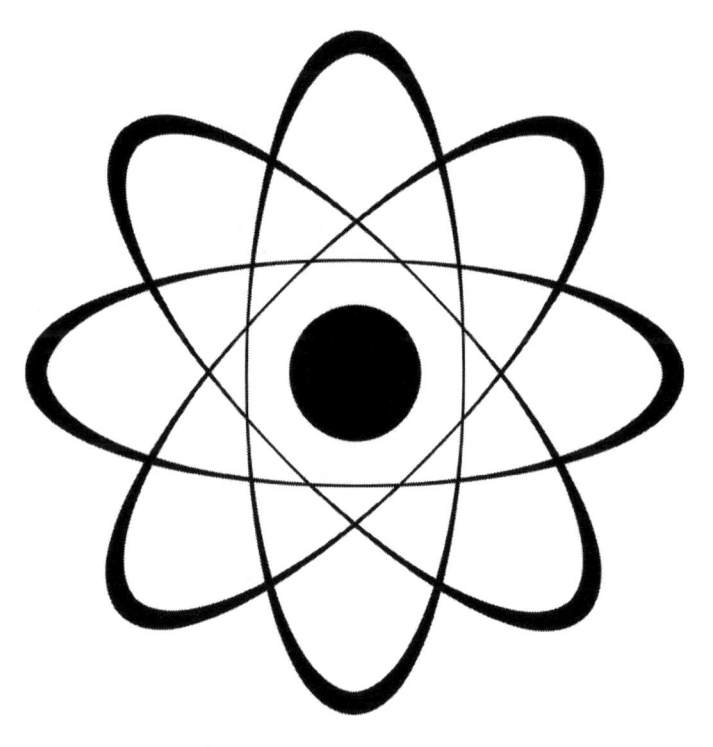

From what we know about quantum physics so far, there are more puzzles than answers. The gist of it all is that you might come across a revelation that sparks a new school of thought. Wouldn't that be awesome?

The 60-Minute Quantum Physics Book

Some of the theories you come across in quantum science might be absurd or unimaginable in application. Take Schrödinger's cat, for example: Imagine a cat that is supposedly dead and alive at the same time. Try explaining that to grandma. (More on this mental exercise later.)

Away from the quirkiness, quantum physics is applicable in many aspects of our lives today. There are many advances in modern society that have been a success due to principles and theories of quantum mechanics. You might not know this, but without quantum physics, there would probably be no personal computers. How is that possible? Simple answer - the transistor. This tiny semiconductor amplifies electronic signals, transferring electrical power from one point to the other. Transistors are present in many electrical appliances that we use every day.

One of the challenges to understanding the application of quantum mechanics in real life is that, in many cases, it contravenes principles we would usually consider to be common sense. Our ideas of realism, locality, and causality are constantly challenged by quantum mechanics.

- **Realism**

The 60-Minute Quantum Physics Book

You are aware of the sun's existence such that, in the darkness of the night, you look forward to the sun's warmth in the morning. In the morning, you don't have to glance at the sun to know it's there.

- **Causality**

If you press the power button on your remote control, the TV turns on. This shows a chain of events where one action leads to another. Causality studies the cause and effect relationship between events.

- **Locality**

This principle states that objects can only be influenced by their immediate surroundings. Therefore, pressing the power button on your remote control will only turn the TV on in the present moment. It cannot turn the TV on a hundred years from the present moment, or even tomorrow.

Most of our lives revolve around these principles. This is how life unfolds before our eyes every day, such that we cannot think of things happening in any other way. Unfortunately, these principles do not always hold in quantum mechanics, hence the awkwardness. Let's briefly look at the impact of quantum mechanics in our lives.

The 60-Minute Quantum Physics Book

Quantum Computing

Quantum computing has been described by many experts as the next frontier in computing. The difference between normal computing and quantum computing is in the way the devices process information. Normal computers process information in the form of bits (binary digits). Quantum computing, on the other hand, uses qubits (quantum bits). Bits and qubits are different in that quantum bits can exist in superimposed states. This means that, before you determine the value of a qubit, it can hold both binary values (0 and 1) simultaneously. Bits, on the other hand, can only assume one value.

Quantum computing is largely still in the experimental phase of development, but if the speeds are anything to go by, these computers will process information at speeds we probably only see in the movies. With top tech giants like Google and NASA leading the global charge into the new age of computing, it is just a matter of time until quantum computing becomes a commercial reality.

The 60-Minute Quantum Physics Book

Cryptography

In life, some concepts seem unfathomable until you benefit from them. One of these is cryptography. Many people don't take the security of their devices and communications seriously until they are breached and have to suffer the consequences. Cryptography is the process of securing communication devices and methods such that the content is only visible to the sender and authorized recipient. In simple terms, we are talking about data encryption. From emails to WhatsApp messages, you use encryption protocols all the time.

Classic cryptography uses encryption keys. Before sending a message, you encode it using a key, while the recipient must use their key to decode the message. This often works, until it doesn't. Even with all the encryption protocols available, loopholes exist in modern cryptography that can be exploited. Encryption keys have been compromised before. Besides, there is always the risk of someone eavesdropping on the communication through any manner of methods - governments and intelligence operatives are notorious for this.

The 60-Minute Quantum Physics Book

Quantum physics proposes a possible solution to guarantee the integrity of communication - quantum key distribution. In this method, the quantum key distribution is an unbreakable key whose identity can only be shared through randomly-polarized photons. You still share information over the usual modes of communication, but you must have a specific quantum key to decrypt the information.

The interesting bit about this is that polarized photons change their state the moment you read them. Any attempt at intercepting or eavesdropping, therefore, will change their polarized state, alerting you of a breach. That being said, since the photons have already changed state, will your encryption key still work? After all, the state to which they matched and could decrypt the information has changed. Intriguing, eh?

The 60-Minute Quantum Physics Book

Chapter 3: String Theory

String theory is a concept that attempts to describe and address different scientific theories, especially relating to atomic-level particles like photons and electrons. You might also come across it referred to as the *theory of everything* on some occasions, given how its theoretical application is universal in application. Of all the areas where string theory has been applied, one of the most important and fascinating one is the impact of gravity on small objects like photons and electrons.

The 60-Minute Quantum Physics Book

Conventional physics has always held that gravity is the reaction of enormous objects like planets to regions of curvature in space. On the other hand, theoretical physicists believe that we should think about gravity as we do magnetism. A magnet will stick to a surface because of the exchange of photons between the magnet and the surface.

Theorists proposed that the behavior of gravity can be explained by eliminating graviton particles (Polchinski, 2005). This is because when the particles collide, the outcome is often infinite and impossible. Strings, however, can collide without such outcomes because they are one-dimensional.

One of the reasons why string theory is referred to as the "theory of everything" is that it changes our interpretations and understanding by replacing all particles of matter and force with strings. The strings do, however, move in complex ways that might even look like the particles we replaced them with. How is this possible? When a string strikes a given note, it obtains similar properties as a photon, depending on the frequency it is vibrating at.

String theory, therefore, is a theoretical framework that describes how one-dimensional strings interact with one another through space. The following elements are key to your understanding of string theory:

Strings and Membranes

In the 1970s when string theory was first developed, scientists believed that the energy filaments (strings) were one-dimensional. One-dimensional, in this case, meant length. This is different from two-dimensional rectangles, for example, that have length and width, or a three-dimensional cube that has length, width, and height.

There were two types of strings involved in this description - open and closed strings. The difference between these two is that the ends of an open string do not intercept, while a closed string is a continuous loop without an open end (Polchinski, 2005). These strings were referred to as Type I strings. Depending on the joining and splitting methods used, Type I strings could go through five different interactions.

Each of the five interactions relied on the string's ability to split or join at the ends. Looking at the description of open and closed strings, we know that you can form a closed string from joining the ends of open strings. Therefore, closed strings are vital in formulating string theory.

Why was this important? By studying closed strings, physicists believed that they had properties that could help in describing gravity from a quantum point of view. Beyond experimenting with string theory on particles of matter, quantum physicists realized that it could be useful in explaining gravity and particle behavior thereof.

Over the years, more scientists contributed to string theory and it became evident that the theory was incomplete without branes (sheets). Individual strings attached at either end through the branes, resulting in a two-dimensional brane.

The 60-Minute Quantum Physics Book

Quantum Gravity

The laws of general relativity and quantum physics have always guided modern scientists. What is interesting about these two laws is that they represent differing schools of thought. General relativity studies the behavior and nature of the universe. This includes the galaxies and planets therein. Quantum physics, on the other hand, studies the smallest objects around.

The consensus, from general knowledge, is that gravity affects all objects on the planet, including the small objects that are subjects of quantum physics. Because of this reason, the laws of general relativity apply to the quantum objects too. Since these two are differing schools of thought, several theories of quantum gravity have been advanced over the years in an attempt to unify them. String theory stands out as one of the most promising of them all.

Unification of Forces

Alongside studies into quantum gravity, string theory also tries to consolidate the important forces in the universe that act on all objects: Gravity, weak and nuclear force, and the electromagnetic force.

Conventional physicists believe that each of these forces is independent and acts on its own principles. Quantum physicists, however, believe that these forces were all but strings interacting with one another in the earliest universe. This was a time when there existed extremely-high energy levels.

Supersymmetry

Each existing particle in the universe can be classified as either a fermion or a boson. Fermions are particles

The 60-Minute Quantum Physics Book

with odd half-integer spins, like 3/2, 1/2, and so on. Most composite particles on earth are fermions, including neutrons and protons. Fermions are known not to co-exist at the same location, in the same state at the same time. Bosons, on the other hand, are particles with integer spins, like 2, 5, 3, and so on. We can identify the nucleus of an atom as boson or fermion by identifying the number of neutrons and protons, and whether they are even or odd, respectively.

In string theory, physicists believe that fermions and bosons share a unique connection known as "supersymmetry". Where supersymmetry is achieved, there must be a fermion for every existing boson. Quantum experiments, however, are yet to detect these particles. Supersymmetry simply represents a mathematical relationship between unique elements in a physics equation. Ideally, the discovery of supersymmetry was independent of string theory. However, incorporating supersymmetry into string theory gave birth to superstring (supersymmetric string) theory.

Supersymmetry is one of the greatest additions to string theory, in that it simplifies equations whose outcomes would generally be inconsistent in the physical world. Because of supersymmetry, insanely-

The 60-Minute Quantum Physics Book

imaginary energy levels and infinite value outcomes are a thing of the past in string theory.

Since scientists are yet to observe fermions and bosons in their experiments, string theory remains a theoretical concept. One possible reason for this is because quantum physicists believe that it takes too much energy to generate fermions and bosons. In the early universe with extreme energy levels, these particles might have existed. However, our universe has cooled down over the years, spreading out most of that energy in the aftermath of the big bang.

Extra Dimensions

While studying string theory, you can expect that the results will only be sensible in a world where at least three space dimensions exist. To put this into perspective, ours is a three-dimensional world (front and back, up and down, left and right). Quantum physicists, however, proposed that there are six additional space dimensions compacted into subatomic sizes such that we can never identify them. Our lives are, therefore, limited to the three-dimensional sheets, and we are unable to access the additional dimensions.

The 60-Minute Quantum Physics Book

Once you understand these concepts, you will find it easier to grasp string theory, and probably dig deeper into the subject. More research and experiments are currently taking place in string theory and other aspects of quantum physics that will certainly be exciting.

Chapter 4: Bohr-Einstein Debates

What is the meaning of "quantum physics"? This question that many scientists struggle to answer convincingly was the bone of contention between Einstein and Niels Bohr back in the day. Their debates essentially challenged the idea of quantum physics as portrayed in the Copenhagen Interpretation. Einstein, particularly took issue with a lot of the assertions held within the Copenhagen Interpretation, more so the notion of a universe

The 60-Minute Quantum Physics Book

independent of observers.

The Copenhagen Interpretation was advanced by Bohr and Werner Heisenberg between 1925 and 1927, attempting to comprehensively explain quantum physics. Since then, there have been several interpretations and explanations proposed by scientists, yet it remains the most-often-discussed interpretation to date.

Today, the general consensus around the debate is that Bohr won the debate, given that he had an answer for all objections raised by Einstein. That being said, however, the debates still persist today, given that several scientists object the concept of an observer-independent universe. As a result, physicists have since proposed several alternatives to the Copenhagen Interpretation, though none of the proposals has received wide acclaim so far, given that they often have several problems making them impossible.

The 60-Minute Quantum Physics Book

Quantum Mechanics Conference

What's interesting about the debate, other than the fame that it has enjoyed over the years, is that of the 29 attendants, 17 had already or would receive Nobel Prizes at some point in their lives. The conference on quantum mechanics was arranged in 1927 in an attempt to reconcile many of the observations in quantum physics that contradicted one another.

Bohr's proposal in the Copenhagen Interpretation was that the position of small entities like electrons could be described using wave equations. However,

The 60-Minute Quantum Physics Book

the entities themselves were non-existent until the moment you searched for them. What Bohr meant, in essence, was that until you looked for something, it did not exist.

Imagine a world where your existence did not count until someone went looking for you. This means that if you are sitting next to someone, they do not really exist until you turn and see them. It is at that moment when you glance and identify them that they exist. Well, like yourself, Einstein was not convinced by this theory.

His rebuttal was that the action of looking at something does not negate the fact that it is what it is. An electron will always remain an electron whether you are looking at it or not. It exists wherever it may be. Einstein and Bohr challenged one another over the years in face-to-face debates or in print.

Einstein's Criticisms

The first issue Einstein had with the Copenhagen Interpretation was on whether quantum mechanics in its form, at that moment, was complete. Through thought experiments, he suggested that momentum and position could be known simultaneously to precision.

In another attempt, Einstein tried to highlight the loss of predictability and absence of causation as proposed in the Copenhagen Interpretation. Bohr was unimpressed by Einstein's experiments and suggestions such that he spoke to each participant in

The 60-Minute Quantum Physics Book

the conference individually, trying to convince them that Einstein's experiment was not true. Bohr later wrote to Einstein with a solution that, surprisingly, relied on the theory of general relativity as advanced by Einstein.

After *losing* to Bohr in his earlier criticisms of the Copenhagen Interpretation of quantum mechanics, Einstein abandoned his attempts to discredit Bohr's work by pointing out inconsistencies in the theories. Instead, he switched to aspects of quantum mechanics that he did not agree with.

This concession meant that, in practical terms, Einstein admitted that determining the value of some incompatible quantities was impossible. Their debates, however, did not end. Next on their discussion was whether the particles had any value, even if they were immeasurable.

As much as they disagreed on several theories and concepts, Bohr and Einstein were subtle in their assertions. In the debates, Einstein always tried to prove inconsistencies in quantum mechanics. Bohr, on the other hand, would counter his claims swiftly. In their final debate, Einstein posed a challenge that sparked modern developments in quantum mechanics as we know it in the 21st Century. He challenged the concept of "entanglement", and all Bohr could do

The 60-Minute Quantum Physics Book

was ignore it altogether.

For what remained of their lives after the debates, most quantum physicists adhered to Bohr's school of thought. Einstein, however, continued working in isolation, trying to unify field theory - a dream which, when he died, remained unrealized.

As strange as these debates and the understanding of the Copenhagen Interpretation were, they are always held in high regard in the purview of quantum physics. Indeed, there have been other interpretations over the years that were ludicrous, but all the quantum physicists seem to agree on the fact that the universe holds more mysteries than we will ever fathom - at least not in our timeline. The universe has a tendency of throwing unimaginable surprising facts at us, leaving us with the uphill task of understanding and making use of them.

Chapter 5: Quantum Entanglement

The concepts of quantum theory and quantum entanglement are often shrouded in mystery, yet they are scientific concepts that should, at the end of the day, have concrete implications and meanings. Entanglement is often misconstrued to be a quantum mechanics problem when, in the real sense, it is not. So, what is it really about?

The idea of entanglement arises in cases where you

The 60-Minute Quantum Physics Book

have incomplete knowledge of the state of two different systems. For example, assume you have two objects, available in two shapes. It is fair to say that the objects are independent, if knowledge of the state of one of the objects does not give you any credible information about the state of the second object. The state of the second object, similarly, should not give you credible information about the shape of the first object.

In a situation where you can derive information about the second object from your knowledge of the first, you can conclude that the two objects are entangled. It is not just about knowing, but the information gained from the first object must improve your knowledge of the second. Knowing the state of one object helps you infer the state of the next object, with certainty.

Quantum entanglement follows the same approach. It purposes that there is no independence between the states. Quantum theory further suggests that states can only be described by wave functions (mathematical objects assigned to them). Wave functions are defined by interesting rules, most of which often introduce complications and ambiguities in the long run. You will encounter such rules at an advanced stage in your journey into quantum physics.

The 60-Minute Quantum Physics Book

What you should know about quantum entanglement, however, is the concept of non-independence that we have elaborated on above.

Practically, independent states in quantum systems are rare. This is because each time any two systems interact, their particles collide, creating correlations between the colliding particles. Therefore, quantum system entanglement is a natural occurrence.

Let's think about molecules, for example. In this case, we'll talk about nuclei and electrons. Molecules usually exist in the lowest possible state of energy. In this state, the nuclei and electrons are entangled. Therefore, we can deduce that the positions of the molecular particles are not independent. When the electrons are jolted into motion, however, the nuclei move, too.

Quantum Complementarity

Initially, we assumed that we had two objects available in two distinct shapes. Further to that, let us now assume that they also exist in two distinct colors. Practically, this new information would mean that the objects can exist in either of the colors. However, for quantum objects, this is far from reality. The fact that one of the objects can exist in different colors and shapes is not a guarantee that it can do so simultaneously. Assuming otherwise would be a common-sense folly, which as much as Einstein suggested that it should be considered as a possibility in physical reality, remains inconsistent in experimental quantum physics.

Why is this so?

Ideally, you can measure the shape of the objects. However, when you measure the shape, you lose all the information you have about the distinct color, or measure the color and lose information about the shape. In quantum theory, we cannot simultaneously measure the color and shape. Insisting on one aspect of physical realism will never capture all the features of the objects. What you must do, therefore, is to consider independent mutually-exclusive views of the features. This way, each view gives you a different, yet valid, assessment of the properties. This is the principle upon which Bohr introduced the concept of "quantum complementarity".

In light of this assessment, quantum theory limits our options when determining the physical features of different properties. To escape contradictions in your outcomes, you must first contend with the fact that any property that you do not measure, does not have to exist. Secondly, you must also note that measuring features is an active process, and in doing so, it changes the system in question.

What we have done so far is to look at how difficult it is to assign an independent state to different objects due to quantum entanglement. With more enhanced protocols, you can introduce the concept of "complementarity" and define objects using the

The 60-Minute Quantum Physics Book

many-worlds concept of quantum theory. What this means is that we can prepare one object in a different color presently, and measure it in a different color at a later time. However, it is impossible to assign a color to the object intermediately. We also realize that the object does not have a definite shape.

It might be difficult to understand quantum complementarity and its influence on quantum entanglement using everyday terms because, in the practical world, this experience is unencountered.

Chapter 6: Schrödinger's Cat

Quantum theory experiments are often complicated. More often, you have to put aside common sense and intuition to embrace quantum ideas that eventually challenge your concept of reality. With more methods and features identified over the years, scientists ordinarily discover new ways of testing quantum theories and delivering intuitive outcomes.

Schrödinger's cat is one experiment that challenges reality. Imagine a life where realities keep shifting from time to time, and more importantly, one where immediate opposites can coexist. In theory, this makes sense, but in practice, the idea itself is confusing, to say the least. For example, how else would you conduct an experiment to compare two contrasting events, when by virtue of quantum physics, they both exist? It's like trying to prove that a *yes* is not a *no* while at the same time, aware that the *yes* can be a *no*.

The Schrödinger cat experiment is a strange one, but one that makes sense in the quantum realm. It involves a cat that is neither alive nor dead.

Superposition

The 60-Minute Quantum Physics Book

Schrödinger's experiment highlighted a common principle in quantum mechanics - that things can be certain, or real, but never both. How does this experiment highlight the absurdities of quantum physics? One of the quantum concepts that were introduced early in the 20th Century was that an object could be two or even more things at the same time. This, in itself, defies everything you know about logic. This idea of co-existing states is what is referred to as "superposition" in quantum mechanics.

Erwin Schrödinger devised a crazy thought experiment with the intention of showcasing the contradicting nature of superposition. How did Schrödinger arrive at this point, using his experiment to show that the cat could be living and dead simultaneously?

In his experiments, Schrödinger showed that all the qualities measurable about a system (the state) can be described using a wave function. Wave functions include all the information about the system that can be held. Each of the wave functions used in Schrödinger's equation was a solution to the equation. However, things got murkier when he suggested the possibility of combining two wave functions to form a third, which obviously carries contradicting information. This is the superposition state. He further added that the number of wave functions that can be combined to form a superposition is endless.

The "Experiment"

Schrödinger published an article in 1935, expressing his disdain over the concept of the collapse of wave functions, measurement and the contradicting states in quantum physics. Unknown to him at the time, this would propel him to plaudits in the history of quantum physics.

Describing his "experiment," he placed a cat inside a steel chamber. After that, he introduced a radioactive substance and observed the effect in an hour. The probability was either atom decay or none. (Keep in mind, this was just a thought experiment - a challenge of the status quo - and not a real experiment on a real cat.)

From this, there was a 50:50 chance that the atom decayed and the cat died at the end of the experiment. If the atom decayed after an hour, the cat would die from cyanide poisoning inside the chamber. If the atom didn't decay, the cat would survive the experiment. From this experiment, the cat and the box would be considered a quantum system.

The wave function in this quantum system, therefore, is in a superposition because it involves the wave function that identifies the live cat and the wave function that identifies a dead cat. Following the principles and guidelines of quantum physics, the cat would, at the end of that experiment, be both dead and alive.

What happens to the cat after the hour-long experiment is done? Well, until the time lapses, we cannot tell for sure whether the cat is alive or dead. We also know that we must peek into the chamber at some point because we must know the outcome. Sneaking a peek into the chamber, however, is by itself an additional measurement of the quantum system. The moment we open the chamber to determine the cat's fate, the wave function collapses.

From another perspective, we might open the chamber and realize that the poisonous substance did not decay. The cyanide bottle is unaffected, and the cat survives. The wave function will still collapse. However, what we cannot tell is how we determined the cat's fate. We might harbor some interpretations, but we cannot tell with certainty how the cat survived.

The 60-Minute Quantum Physics Book

The Many-Worlds Interpretation

So far, all we have done is to give a brisk illustration of quantum mechanics under the Copenhagen Interpretation. This is the most widely-used interpretation, so using it to set the foundation of our understanding makes sense. Moving on, what the Copenhagen Interpretation does is to describe a system independent of values - until that point where we assign a value or record a measurement. In the case of Schrödinger's cat, this happens when we admit that the cat is alive.

But, an interesting question arises - what if instead of collapsing, the wave function keeps growing? The result is a relative state, a formulation that was later known as the *many-worlds interpretation*. In this interpretation, the wave function expands instead of collapsing when we open the chamber. The cat is still in a superposition. However, we also become part of the superposition. Our universe also becomes part of the superposition. In essence, we become part of the new system.

According to the many-worlds interpretation, our aim when opening the chamber is not to determine whether the cat is alive or not, but instead, to identify the universe we are in - one where the cat died, or one where the cat survived. Ourselves, and our worlds, therefore, become part of the wave function. Hence, our new universe (whichever it is) is in superposition with the former.

Chapter 7: Double-Slit Experiment

The complexity of quantum physics is more often highlighted through the double-slit experiment. The double-slit experiment is one of the theories of quantum physics whose results have proven strange not just for non-scientists, but also for scientists to date. In this experiment, light is shone on a barrier with two narrow openings. The researcher observes the pattern of interference produced on the observation screen.

The 60-Minute Quantum Physics Book

From our knowledge of general and quantum physics at this point, we are aware of two properties of light: Light travels in waves (classical physics). Light can also travel in photons (quantum physics). Assuming the double-slit experiment is sensitive, the result is a grainy pattern of interference, showing an individual photon on the screen. What we can tell from this is that, while the photon is not divisible, it travels by interfering with neighboring photons. This manner of reverberation is characteristic of matter.

This concept might seem straightforward, yet its interpretation has often proven quite divisive over the years. The Copenhagen Interpretation is widely accepted and supposes that the nature and behavior of photons in this experiment is because the experiment determines what you observe. This is why photons will at times appear like waves, and in some cases, they appear like particles.

As you can imagine, the Copenhagen Interpretation was not convincing to many scientists, with Einstein being one of the most vocal opponents of this approach. Over the years, many scientists have proposed alternate interpretations of the double-slit experiment. As simple as the experiment is in practice, the conflicting interpretations made it one of

The 60-Minute Quantum Physics Book

the most difficult experiments in quantum physics to date.

The interesting thing about quantum physics is that, because of the complexity, many experts treat it as a conundrum. This is because, while the predictions available through quantum mechanics might be useful, attempting to understand why the predictions work as they do will usually drive you up the wall, confusing you more than it provides answers.

Why is this experiment important?

Why would anyone go through all the trouble after all? If the likes of Einstein never succeeded in demystifying the double-slit experiment, why should you dig deeper into it? Well, this experiment is key to understanding the structure and nature of quantum mechanics. That's the fundamental reason for wanting to know more about it.

The Schrödinger equation, also referred to as the "wave equation", is at the heart of understanding quantum mechanics. Like several math equations used in physics, this equation requires different parameters to define a physical state, and find the solution to the equation. The solutions in the double-slit experiment are wave functions.

The 60-Minute Quantum Physics Book

Wave functions tell you the system's state, which in this case, could be one or more atoms, electrons, photons, or any other entity that make up the state. The state, on the other hand, can only give us the probability that the system in question attains a given position.

Beyond the scope of quantum physics, probabilities are a statistical method used to describe different things in day-to-day activities. For example, when rolling a die, the probability that the number 5 comes up, or the probability of employees calling in sick ahead of a holiday that falls on a Friday. Probabilities assume that there is never complete certainty that an event will happen. Wave functions use statistical information to explain what happens in an individual system. The assumption is that, even if you perform an experiment on one photon, you can always expect more than one possible outcome.

The Value of Uncertainty

If there is something you will notice about quantum experiments, it is the fact that there is always an element of uncertainty involved. Naturally, there are no perfect experiments. For this reason, therefore, some uncertainty is expected when performing experiments. However, when it comes to quantum

The 60-Minute Quantum Physics Book

physics, the idea of uncertainty is stretched to another level. Even in situations where experiments are performed with perfect equipment, there exists a limit to the effectiveness of measurements.

In the quantum realm, this kind of uncertainty is associated with the wave-like nature of light and matter. For example, if you study a water wave ripple across a water body, can you tell with certainty the speed of the wave in motion? Can you tell the exact position of the wave at any given point in time?

There is no clear answer because the wave assumes a limited amount of space, and while in motion, it might overlap with other waves, making it impossible to identify each wave. Apart from that, as you watch the wave moving across the water, you must also be aware of the fact that different crests and troughs move at different speeds.

The 60-Minute Quantum Physics Book

It is safe to say, therefore, that the most feasible way of determining the momentum and position of the wave is by averaging a spread of values - hence the uncertainty. Uncertainty in quantum physics, therefore, does not carry the same connotation as the linguistic concept of doubt, but indeterminacy. You cannot be sure what the exact values are, so you work with informed guesses.

According to the uncertainty principle, as advanced by Heisenberg, the minimum-accepted uncertainty for quantum waves must be larger than the uncertainty in the momentum, but smaller than the uncertainty in the position. Taking this back to the double-slit experiment, the wavelength is a function of the momentum. Therefore, the interference pattern observed becomes a measure of the momentum. This also means that by determining the slit through which the photon passed, which, in essence, measures the position of the photon, the uncertainty increases. It might be possible to identify where a photon lands, but identifying the path it followed to arrive at the destination is impossible.

The 60-Minute Quantum Physics Book

Chapter 8: Teleportation

Forget about everything you have seen in the movies - at least that's the closest most of us have come to

The 60-Minute Quantum Physics Book

understanding or experiencing teleportation. Teleportation has baffled many scientists and enthusiasts alike. It is one of those mysteries that quantum physics attempts to solve, proving it more science than fiction.

Quantum physicists have since discovered the possibility of transporting a quantum state over a relatively-long distance without sending the quantum state itself. Quantum teleportation refers to a process where quantum information can be transferred from one place to another, using traditional communication methods and quantum entanglement between the sender and recipient. Other than what we see in the movies, scientists have proven teleportation a success in recent years.

Chinese and Austrian scientists have recently teleported three-dimensional quantum states - a huge milestone in the history and future of quantum computing (University of Vienna, 2019). Before this, quantum teleportation was only possible in theory. The researchers managed to teleport the quantum state of one photon to a distant state. Before their experiment, it was only possible to transmit two-dimensional qubits with values 0 or 1. The researchers, however, managed to teleport a qutrit (third-level state).

The 60-Minute Quantum Physics Book

It is important to mention that, in quantum mechanics, the binary values 0 and 1 can exist simultaneously, unlike in normal science where only one of them can exist at a time. It is also possible for these values to exist in a state in between their simultaneous existence. It is based on this approach that the researchers teleported a third possible state - 2.

The advancement above moved the concept of "teleportation" strides ahead, further away from science-fiction and closer to reality. It might not be possible to practically teleport people or objects, but in theory, science has proven that teleportation is feasible. The idea of teleportation we are looking at is a philosophical one, where you scan an object or state

The 60-Minute Quantum Physics Book

and transmit the scanned information elsewhere, using that information to build a new object or state.

This scanning-and-reassembling approach to teleportation is possible under quantum entanglement, which we discussed earlier. The concept of "quantum entanglement" applies when at least two particles exist in mutually-exclusive states. When this happens, it follows that determining one state automatically determines the other.

Think about this like ordering two different types of pizza of the same size: Once the delivery person arrives, you cannot tell which flavor is in either of the boxes. However, once you open the first box, you immediately know what flavor is in the second box without seeing it. It doesn't matter where the second box is in your house relative to the first one.

This concept applies to quantum particles too. Assuming you have two entangled particles, knowledge of the state of one of them automatically tells you the state of the second particle, and like our pizza example, the state is independent of the distance. This is why quantum entanglement is important in quantum teleportation.

That's as simple as quantum teleportation gets - or rather, the simplest explanation of why it is possible.

The 60-Minute Quantum Physics Book

If you entangle two particles and send them to a distant location, we can, in theory, use the principle of quantum entanglement to teleport anything between the particles. To teleport any object, all you have to do is ensure it is included in the entanglement. The process of making this happen is where all the math wizardry takes place.

Quantum teleportation is still largely a concept. We cannot expect to teleport large objects yet. Even where teleportation is achieved, it is also not possible to keep the particles entangled over a long duration or a long distance. Further, the technology in use today cannot allow the teleportation of particles larger than a few atoms. The teleportation experiment, however, has been performed successfully by many scientists over the years - a glimmer of hope that you might teleport to your favorite holiday spot in the future. Goodbye, expensive flights!

Is Teleportation Practical?

The earliest papers on teleportation were purely theoretical. They followed quantum mechanics guidelines and mathematical rules to predict teleportation. Several experiments have since been advanced, demonstrating the feasibility of quantum teleportation.

The 60-Minute Quantum Physics Book

Before teleporting a quantum state, it must be encoded within the possible paths that the photon will follow. These paths are encoded as three optical fibers. It is important to mention that it is possible to identify the photon in each of the three optical fibers simultaneously.

Teleporting the three-dimensional quantum state introduces a different method using a multiport-beam splitter. The splitter connects all the optical fibers together by directing the photons through various inputs and outputs. You also introduce auxiliary photons into the beam splitter. Take note that auxiliary photons might interfere with the objective photons. What we are interested in, here, is the interference pattern.

By choosing specific interference patterns, it is

possible to transfer quantum information from the input photon to another photon some distance away from it, without any of the photons coming into physical contact. What's more interesting about this concept is that this teleportation concept is not restricted, and can apply to three or more dimensions.

Quantum teleportation is generally one of the best illustrations of quantum entanglement, or what Einstein referred to as *spooky action at a distance*. In quantum entanglement, particle properties of one state affect those of other particles even when they are some distance apart. Teleportation, therefore, is about two entangled particles some distance apart, such that a third particle teleports its state instantly to the two aforementioned particles.

Chapter 9: Interesting Quantum Physics Facts

From the very beginning, quantum physics can be intimidating. The quirkiness in it baffles even some of the top physicists in the field. Studying the tiniest possible units of any entity does come with its challenges. At times, it might even feel like the discipline is more theory than application. Quantum physics is, however, not incomprehensible. Once you understand the important concepts that guide scientists in the field, you will find it a lot easier to understand.

Without these concepts, most people think about

issues from the perspective of the normal world, and that's where things get fuzzy. It's like magic at times; quantum physics makes you step away from conventional realities and see things from a different perspective. Let's discuss some interesting facts that you might come across:

The Entanglement Concept

The theory of quantum entanglement might as well leave your brain cells entangled. In conventional physics, nothing moves faster than the speed of light. Quantum entanglement, however, suggests otherwise. In quantum entanglement, you learn that tiny particles reverberate at speeds faster than the speed of light. In this case, none of the particles acts independently of the others - hence they cannot be described as such.

Over the years, scientists have suggested several theories to explain the theory of quantum entanglement. If you find it crazy, well, it probably is, and Einstein thought so too. He even coined the phrase *spooky action at a distance*, to express his sentiments about the theory of quantum entanglement, as also mentioned in the previous chapter.

The 60-Minute Quantum Physics Book

Building on this, another interesting feat you will learn in quantum physics is that everything is about probabilities. Throughout history, one thing quantum scientists agree on is that predicting the outcome of an experiment with utmost certainty is impossible. Scientists always try to predict the probability of arriving at a possible outcome. There is, therefore, a variance when it comes to describing a quantum state, such that the mathematical description will always appear in the form of a wave function.

Virtual Particles

When you walk into an empty room, naturally, you can visually see that it is empty. In quantum physics, however, that room is not empty. What you see as empty is a room filled with matter, antimatter, and energy. Because of the energy within the room, random particles keep vibrating in and out of existence.

In quantum physics, it is believed that the random particles are locked in a cycle where they appear, bump into one another and disappear. All this takes place within a billionth of a second - such a short time that you would never recognize it. This is one of the quirky theories in quantum physics that scientists are still exploring. You have to admit that this probably changes your perspective of "empty" spaces.

The Weight-Speed Argument

Here's an interesting one - your weight is heavier when you move faster. Take a moment and think about it. Does that even make sense? Does it mean that your body weight varies throughout the day?

In quantum physics, the theory of relativity, advanced by Einstein, suggests that you might weigh more when you are moving faster. We learn that energy and mass are similar, and as such, they are interoperable. Assuming that you exert more energy on an object moving at super speed, it would be slower. At the same time, the extra energy is necessary to sustain the speed while managing the added mass on the object.

The 60-Minute Quantum Physics Book

Sounds Like Magic

There is so much that goes on in quantum physics that will challenge your belief in the normal things that happen in the physical world. Things like teleportation, for example, are unfathomable, apart from what we see at the movies. For most people, you can even misconstrue most of what happens in quantum physics to be magic. Make no mistake, magic is the last thing you should associate with quantum physics.

As strange as the things you can predict with quantum physics are, they all fall under mathematical principles and rules that have existed since the beginning of knowledge. Therefore, if you come across a quantum idea that sounds too strange and almost impossible, there's a good chance it's true. Most of the things you can do in quantum physics are viable within the boundaries of thermodynamics. However, when it comes to common sense, that's a different story altogether.

The 60-Minute Quantum Physics Book

The Thrills of Quantum Theory

In essence, quantum theory describes the behavior of energy and particles, albeit at the tiniest scale. It also supposes that particles can exist in more than one place at the same time, or even pass through a wall. Quantum physicists have also shown in the past that you can link pairs of particles even if they are in different sections of the room or different parts of the universe. This is covered under "quantum entanglement".

It might seem like scientists have finally figured things out, given that all these theories and concepts in quantum science are understandable. However, quantum theory is more frustrating than ever. Some scientists even believe that the moment you think you understand quantum physics, you probably have lost the plot.

So, what's thrilling about it? In simple terms, the thrill is in the fact that the experiments work. Explaining how or why they work might be another story altogether. Most experiments in quantum physics attempt to verify whether the quantum predictions made prior are accurate or not.

The 60-Minute Quantum Physics Book

Discoveries about the behavior of photons and electrons have led to amazing developments with the likes of transistors relevant in smartphones and computers. However, developers who create these devices often follow rules whose complete understanding they do not possess. That's the interesting thing about quantum physics.

Some scientists think of it as a cocktail. You might have all the necessary ingredients to make an amazing cocktail. You can even give a step-by-step brief on how to mix the ingredients correctly and in the right portions. What you cannot do, however, is explain what happens to each ingredient, and how they interact with one another to deliver that precise cocktail taste.

The discussions above show not just how strange quantum physics is, but also prove that it's an exciting field of study for someone who loves the thrill of seeking the unknown.

The 60-Minute Quantum Physics Book

Conclusion

Of all the areas of study in physics, none has enjoyed more attention in recent years than quantum physics. As complex and confusing as many find it, it still draws plaudits from physicists, other scientists, and non-scientists alike. Perhaps it is the weirdness or the fact that the supposed ambiguity challenges theories and concepts that everyone considers as the norm.

Quantum physics is an exciting field of study if you are the kind of person who enjoys the thrill of solving puzzles. You will come across many problems that demand unconventional approaches to solve. For an out-of-the-box thinker, this should tickle your fancy. Quantum physics is also a good place to be for the curious mind that asks questions often.

There have been many developments in quantum physics over the years, yielding several sub-categories. This is good because it keeps things exciting. With time, you will realize that it is possible to gain expert knowledge in one quantum field, but be average in another. Due to the interconnectivity of quantum disciplines, you can always learn from experts in any one area.

The 60-Minute Quantum Physics Book

Quantum research is a promising field that is expected to transform industry and science over the years. From predictive algorithms to financial data analysis and developments in science, there is so much to look forward to when it comes to quantum research. The goal of quantum research transcends supercomputers. We are talking about information science, technology, metallurgy, and materials - to name a few.

Quantum supremacy is feasible in the near future. To get there, however, you must start from scratch. We cannot be oblivious to the fact that quantum physics involves a lot of math. For the avoidance of doubt, we are not talking elementary-level math here, but insane, almost apocalyptic-level, math. What does this mean for someone who is not the biggest math enthusiast? You don't have to fret, and let's explain why:

Remember years back, when computer programming was the preserve of geeks? Programming has evolved so much that today, such that you can learn how to code without prior knowledge of programming. Today, there are tutorials all over the Internet, and depending on your dedication, you can learn how to code in a few weeks or even days.

The 60-Minute Quantum Physics Book

So, back to quantum physics. We find ourselves in the same spot. The math might be intimidating right now, but as you can see from what we have discussed in this book, the concepts are intriguing, and you can easily understand them. Knowledge of the basics should help you build your interest in quantum physics. With time, even the math will be easier and more relatable for most people.

The best thing about quantum physics is that, at this point in time, the field has developed so much and you can choose a quantum discipline that suits you. Give the advancements in technology, many would gladly venture into quantum computing. Considering how much you can leverage your knowledge of computing in the global markets today, there are incredible options available for you.

What we must not forget, however, is that, at the end of the day, you must prioritize the science aspect in whichever quantum discipline you pursue. This sets the backdrop upon which future transformative industrial and scientific progress can emerge and thrive.

The 60-Minute Quantum Physics Book

Bluesource And Friends

This book is brought to you by Bluesource And Friends, a happy book publishing company.

Our motto is **"Happiness Within Pages"**

We promise to deliver amazing value to readers with our books.

We also appreciate honest book reviews from our readers.

Connect with us on our Facebook page www.facebook.com/bluesourceandfriends and stay tuned to our latest book promotions and free giveaways.

References

Anon. (2016). String Theory in Cosmology. *International Journal of Science and Research (IJSR)*, *5*(6), 1821–1822. https://doi.org/10.21275/v5i6.nov164664

Bates, D. R. (1962). *Quantum Theory;* New York, London: Academic Press.

Bernstein, J. (1966). Einstein and Bohr: A Debate. *The Physics Teacher*, *4*(6), 258–265. https://doi.org/10.1119/1.2350996

Bouwmeester, D., Ekert, A. K., & Zeilinger, A. (2011). *The Physics of Quantum Information: Quantum Cryptography, Quantum Teleportation, Quantum Computation.* Springer.

Clegg, B. (2009). *The God Effect: Quantum Entanglement, Science's Strangest Phenomenon.* St. Martin's Griffin.

Clerk, A. (2012). Seeing the "Quantum" in Quantum Zero-Point Fluctuations. *Physics*, *5*. https://doi.org/10.1103/physics.5.8

De Martini, F., Mussi, V., & Bovino, F. (2000). Schrödinger Cat States and Optimum Universal Quantum Cloning by Entangled Parametric Amplification. *Optics Communications*, *179*(1–6), 581–589. https://doi.org/10.1016/s0030-4018(00)00611-8

Forrester, R. (2018). The Bohr and Einstein Debate:- Copenhagen Interpretation Challenged. *SSRN Electronic Journal.* https://doi.org/10.2139/ssrn.3221514

Gerardi, S. (2018). Quantum Superposition/Social Superposition and Classic Sociological Theory. *Sociology Mind, 08*(01), 21–24. https://doi.org/10.4236/sm.2018.81002

Gold, A. (1986). Phonon Effects on the Transport Properties in the Fractionally Quantized Hall Regime. *Physical Review B, 33*(8), 5959–5960. https://doi.org/10.1103/physrevb.33.5959

Hiton, L. (2017). *The Theory of Relativity.* Cavendish Square.

Mcevoy, J. P., Zarate, O., & Appignanesi, R. (2001). *Quantum Theory.* Icon Books/Totem Books.

Paul, A., & Qureshi, T. (2018). Biphoton Interference in a Double-Slit Experiment. *Quanta, 7*(1), 1. https://doi.org/10.12743/quanta.v7i1.67

Polchinski, J. G. (2005). *String Theory. Vol. 2, Superstring Theory and Beyond.* Cambridge University Press.

Posiewnik, A., & Pykacz, J. (1988). Double-slit Experiment, Copenhagen, Neo-copenhagen and Stochastic Interpretation of Quantum Mechanics. *Physics Letters A, 128*(1–2), 5–8. https://doi.org/10.1016/0375-9601(88)91032-8

Reece, M. (2014). Tracking Subatomic Physicists. *Science, 343*(6178), 1434–1434. https://doi.org/10.1126/science.1251659

The 60-Minute Quantum Physics Book

Rogalski, M. S., & Palmer, S. B. (1999). *Quantum Physics*. Gordon And Breach Science Publishers.

Science X Staff. (2019, August 23). *Complex Quantum Teleportation Achieved for the First Time*. Phys.Org; Phys.org. https://phys.org/news/2019-08-complex-quantum-teleportation.html

Squires, G. L. (2018). Quantum Mechanics | Definition, Development, & Equations. In *Encyclopædia Britannica*. https://www.britannica.com/science/quantum-mechanics-physics

Wilce, A. (2019). Conjugates, Filters and Quantum Mechanics. *Quantum*, 3, 158. https://doi.org/10.22331/q-2019-07-08-158

Wilson, A. (2013). Demystifying the Multiverse. *Metascience*, 22(3), 583–586. https://doi.org/10.1007/s11016-013-9759-5

Yousif, M. (2016). The Double Slit Experiment-Explained. *Journal of Physical Mathematics*, 7(2). https://doi.org/10.4172/2090-0902.1000179.

Printed in Great Britain
by Amazon